OILSEED RAPE AND BEES

OILSEED RAPE AND BEES

by

ALLAN CALDER

with photographs by
Ted Hooper, Michael Singleton
and Paul Taylor

NORTHERN BEE BOOKS
Mytholmroyd : Hebden Bridge

© Allan Calder 1986
First published April 1986
by Northern Bee Books, Scout Bottom Farm, Mytholmroyd,
Hebden Bridge, West Yorkshire
Acknowledgements are made with thanks to the following:
Cover design and photograph by Paul Taylor
Text set in 11/13 Bembo by G. Beard & Son Ltd., Brighton
Printed by Tony Ward at the
Arc and Throstle Press, Todmorden, Lancs.
Colour Printing by G. Beard & Son Ltd., Brighton

British Library Cataloguing in Publication Data

Calder, Allan
Oilseed rape and bees.
1. Rape (Plant)—England
2. Pollination by insects—England
3. Honeybee—England
I. Title
633.8'53   SB299.R2

ISBN 0-907908-35-7

Contents

Illustrations

PHOTOGRAPHS

LINE DRAWINGS

Foreword

This short account is an attempt to draw together, from a number of scattered sources, information for beekeepers about oilseed rape and its relationship to bees and bee-keeping. It is based on authoritative accounts in various special fields, all of which are identified at suitable places in the text.

A special debt is owed to particular individuals with extensive practical experience who generously gave of their time and expertise in describing and discussing the opportunities and problems which oilseed rape presents to the apiarist. They are Rev. F. Cordingley, Malton; Brian Eade, Mountain Grey Apiaries, Goole; W. S. Robson Jun., Chain Bridge Honey Farm, Horncliffe, Berwickshire (in his lecture at the Inter-College Conference, Craibstone, Aberdeen on September 1, 1984); Barry Potter, Senior Lecturer in Horticulture and lecturer in beekeeping, Askham Bryan College of Agriculture, York and members of the staff of A.I.S. (Brown Butlin) Ltd., Selby. The latter sections, 'From the apiarists viewpoint' and 'Bees and Crop Protection' rely almost entirely upon their contributions.

The occasional speculations (and any errors) are the sole responsibility of the author.

A remarkable history

One of the attractions of beekeeping is that it is work against the odds.

The odds are expressed in the nature of the primary resources which every British beekeeper seeks to exploit: a capricious climate, a varied, unpredictable plant environment and an energetic, unpredictable and dangerous insect species. Starting with such raw materials it is the objective of every act of bee management to regularize apparatus and control events so that the effects of unpredictability are reduced and the likelihood of getting a useful honey crop increased. However, none of the three natural systems – the bee community, the plant population, or the climate – are well understood and the beekeeper must repeatedly attempt to find (and put his trust in) his own individual solutions to a host of particular local problems though he can seldom, in the present state of his art, be sure he is right. Were it not for all the fun to be had out of argument and discussion, he might wish for repetitive good weather, beestocks which always behaved in the same way and beeplants boringly consistent in their yield of pollen and nectar. Beekeeping conferences and club meetings, and the controversialism which always attends them, show how vigorously the beekeeping community seeks reliable information upon which to base its view of the bee world and its actions to exploit it.

There is however, one fraction of the beekeeper's resources which, since about 1950, has grown dramatically, whose dependability is outstanding and whose future, in spite of initial controversy and doubt, seems increasingly assured. Insect-pollinated field crops grown for their seed were not widespread in Britain before the Second World War. Field beans were well known though the acreage was not extensive. But a new crop now grows in the arable fields and has increased so explosively as to alter the appearance of the country landscape during May and June each year. It is oilseed rape, a winter surviving variant of the botanical species *Brassica napus*, near relative of a group of other plants much prized by man such as cabbage, kale, cauliflower and sprout.

As a source of nectar it is unequalled in size or in concentration, except perhaps by heather. As one experienced Yorkshire beekeeper put it, 'There aren't enough bees in Europe to get all the honey out of Yorkshire rape'. Its agricultural success is due to strong demand for two bi-products of the seed of the plant: oil, which can be squeezed out by seed-crushing machinery, and seed-meal, the valuable residue. Both the products and the plant have striking histories and with their future important interests in British and European beekeeping are bound up.

The plant itself or its near relatives, appear to have been written and talked about for some 4,000 years. Rape seed is mentioned in the writings of ancient India and in Japanese literature of 2,000 years ago. For many centuries the crop appears to have been grown, as a source of oil, in countries where olives and poppies (alternative oil sources) do not thrive and, from the 13th century, rape oil was an important means of illumination in Northern Europe until replaced by petroleum (27).

Had the light from burning oil been the only benefit from the crop its history would have ben short but it proved to have other virtues.

The extensive operations undertaken in Cromwell's time to drain parts of the English Fens and to turn the marshes into arable land and pasture were under the direction of Dutch engineers who knew from their work on the European continent, of the value of rape seed as a 'pioneer crop' in land reclamation (6).

The beginning of their work is signalled by a governmental Act of 1649. It has the following title and preamble:

'An act for drayning the Great Level of the Fens . . . .
Whereas the said Great Level by reason of frequent
overflowing of the rivers Welland, Neame, Grant,
Ouse, Brandon, Mildenhal and Stoak have been of
small and uncertain profit, but, (if drained) may be
improved and made profitable, and of Great Advantage
to the Commonwealth, and the particular Owners,

2

Commoners and Inhabitants, and be fit to bear Cole-seed and Rape-seed in great abundance, which is of singular use to make Sope and Oyls within this nation, to the advancement of the trade of Clothing and Spinning of Wooll, and much of it will be improved into good pasture for the feeding of Cattel, and for Tillage, to be sown with Corn and Grain, and for Hemp and Flax in great quantity, for making all sorts of Linen Cloth, and Cordage for shipping and Trading at home and abroad, will relieve the Poor, by setting them on work, and will many other ways redound to the great advantage and strengthening of the Nation . . . .' (16).

The role of oilseed rape as a pioneer crop, a colonizer of virgin land and a harbinger of new wealth could hardly be put more forcefully and, during the next two hundred years it became widely grown in England. A flourishing oilseed-crushing and oil-extracting industry developed simultaneously (it is still strongly represented today in Hull, for example) and what might be called the first oilseed boom strengthened and continued until about 1850. However, it finally subsided, and the field rape crops with it, because of the growing competition from imported oilseeds from Russia and the tropics. It was to revive in a hundred years, but not before two world wars had intervened, each of them impressing on Britain the vulnerability of her sea-routes and her dependence upon other countries for many important industrial materials. Advantages in the development of a home produced oil supply became obvious (6).

Moreover, in the 1960's the western world took leave of butter and turned to margarine as a source of domestic fat. It was proclaimed that a diet rich in animal fats ('saturated' fats, as they are called) increased the likelihood of heart attack and that to eat margarine, with its high complement of 'poly-unsaturates', did not carry that risk. Poly-unsaturated fats are found in vegetable oils such as rape oil and the demand for them was to rise rapidly with the increasing popularity of the various high-street commercial brands of soft margarine.

It happened that from 1950 an increasing number of southern English farmers had begun to use rape as a 'break' amongst the successive cereal crops of intensive barley growing. Because of its agricultural virtues and because of the price paid at the oilmills for harvested seed the practice had spread and at least one group of growers (Wessex Agricultural Producers, now United Oilseed Limited) set about the organized promotion and marketing of the crop. Fortune, and changes in the dietary preferences of Western nations appear to have favoured their efforts for, within ten years and with the entry of the United Kingdom into the European Common Market (1973), the price of rapeseed climbed steadily to a high level and the almost nationwide spread of the crop had begun (6, 13).

Between 1968 and 1972 about 6,000 hectares (15,000 acres) of rape were grown in England annually, largely in the South Eastern Region, the East Midlands and East Anglia. In the space of ten years, to 1983, and in spite of a universal interruption in progress during the 1970's, the area had increased some thirty-fold and many thousands of hectares of arable land in Yorkshire, Humberside and further north had been put down to the crop (6). The large-scale invasion of the Borders and Scotland was to follow.

From the British farmer's viewpoint oilseed rape has numerous advantages. It has proved remarkably well adapted to the U.K. climate and, with its strong taproot system, will grow on light soils provided they are deep, as well as on heavier where it does best. Because autumn-sown wheat, barley and rape have comparable growth cycles the farmer can introduce the latter as a crop alternative to either of the others and so make an advantageous 'break' in the succession of cycles of cereal. Rape can be harvested with the same machinery as is used for getting in cereal so that little or nothing need be spent on special equipment for the rape. If potatoes or sugar beet are part of a current rotation their replacement by rape allows the expensive potato or beet harvesting machinery to be sold with resultant reduction in overhead expenses. In growing rape the farmer need make no additional investment in livestock as he would presumably be obliged to do if he were to grow grass as a 'break' crop (8).

Though no threat to cereals as a carrier of disease, oilseed rape is itself susceptible to certain pests and 'soil-

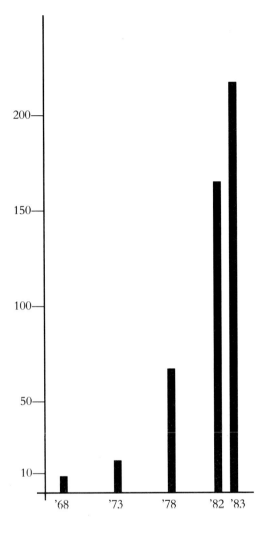

Fig. 1.   Area under oilseed rape in England, 1968 – 1983 (in 1000 hectares). Based on figures given in Bunting, E. S., 1984.

borne' diseases – diseases capable of lying dormant in soil during a period of years and re-infecting their preferred plants should those again be grown before the disease organisms die out. As a result it is recommended that rape should not be sown on the same land oftener than 1 year in 5 and, because live disease particles can be blown about by the

5

wind, splashed up in heavy rain and be carried about by animals and machinery, areas sown with rape should not overlap nor touch one another in successive years but be separated by safe distances (8).

However, the decisive factor from the farmer's angle in favour of oilseed rape is the financial benefit coming to him from the European Economic Community through 'intervention' (the use by governments of public money to increase the rewards paid to producers of selected commodities). E.E.C. intervention emerges as an attractive price paid by oil-millers for the harvested seed and is, at present, enough to make rape a more profitable crop than, for example, wheat.

Of this massive resource, spread widely over huge acreages in southern, eastern and northern Britain and popular and profitable to the farmer, the beekeeper is a part-inheritor. For him every wide-stretching yellow field is a crock of gold. Supposing an average potential honey yield of about 30 lbs. per colony per acre, one expert estimate of the worth of the honey crop from the acerage of 1981 was £7.5 million! (51). By 1984 it could well have approached twice that value and, so long as the rape continues to be grown, so long will this opportunity continue for the beekeeper.

The future of the yearly rapeseed crop depends upon many factors but important among them are:

1. the demand for rapeseed oil and the popularity of commercial products derived from it

2. the demand for rapeseed meal and the success of animal feedstuffs incorporating it

3. the success of its competitors: soyabean, oilpalm, sunflower, peanut, maize

4. the social effects of political support for the crop and its growers (e.g. by 'intervention' or subsidy).

Just as in olden times there were both industrial and domestic uses for the oil from rapeseed, so there are still. Today industry requires it for the production of soap, synthetic fibres, plasticizers, water repellants, waxes and surface-active substances. It is still used as a lubricant. At home it appears in the kitchen as margarine, cooking fats, cooking oils and salad dressings (3, 41).

The upsurge of interest, during postwar decades, in the adverse effects of animal fat on cardiac health focused interest on alternative sources of oily substances with chemical characteristics different from those in dairy- or butchershop-products. Vegetable oils, rich in the now famous 'unsaturated fatty acids' or 'poly-unsaturates' became widely used in the manufacture of imitations of butter: the various high street brands of margarine. But the expansion of rapeseed oil into this new consumer market was hindered because scientists showed that it contained a substance apparently causing disorders in the very organ (the heart – albeit the heart of rats) that the changeover from butter to margarine had been meant to protect. The presence of this substance, known as erucic (pron: eroosic) acid, caused serious concern and was responsible for an interruption, during the 1970's, in the steady adoption of the rape crop and its products but the situation was saved by the work of plant breaders who were able, in the course of about ten years, to produce new varieties of the rape plant free from erucic acid in the seeds. The way was then open for the full exploitation of the oil from rapeseed in the manufacture of wholesome human food (4, 27).

Artificial feedstuffs for farm animals are an important modern agricultural resource and the oil-cake residue from the milling and pressing of oilseeds is an indespensible ingredient. The oil-cake from the milling of rapeseed proved to contain a store of protein comparable to the best of other oilseeds such as soya bean and the prospect of selling it to feed-manufacturers further increased the potential value of the rape crop (15). But serious difficulties emerged when it was discovered that a diet of rapeseed meal had disastrous results for many animals, including poultry and pigs. The work of important glands (the thyroid and the pituitary) was sabotaged by substances in the meal. Bone deformities and liver damage appeared in poultry and the quality of eggs was spoiled. Adult pigs showed a loss of apetite and a decline in their rate of growth. The number of piglets born to pregnant sows was reduced when the latter were fed rapeseed meal (3, 5, 7, 41, 45).

The offending substances in the meal turned out to be complex chemicals known as glucosinolates and their pre-

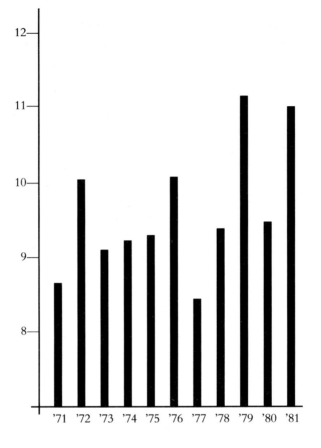

Fig. 2. Worldwide area under cultivation for rapeseed, 1971 – 1981 (in million hectares). Based on figures given in Kramer, Sauer and Wallace, 1983. Compare production before and after 1979.

sence represented an important loss to the industry since, at best, only small quantities (about 15% by weight) of rapeseed meal could safely be incorporated in feedstuffs.

Again, however, the situation was transformed by the remarkable work of the plant breeders. They had already altered the rape plant so that it produced only negligible quantities of erucic acid. Now their modern skills enabled them to limit also its ability to manufacture the troublesome glucosinolates. Though this proved to be an extremely difficult problem new varieties of oilseed rape with both these desirable properties were successfully bred during a number of years of work and commercial quantities first grown in Canada in 1974. So important was the change that the Canadian industry now distinguishes the oil and the meal from these specially bred varieties of the rape plant by a particular name, 'Canola'. Canola oil is guaranteed to be low in erucic acid and the canola meal low in glucosinolates. Rape varieties producing both oil and meal with these qualities are called 'double-low'. Plant breeding with similar objectives has also been going on in Europe and the first European double-low variety was released into commerce in 1984. The potential demand for meal from this and any other similar varieties is likely to treble (6, 27).

The advent of 'double-low' varieties in Europe means that there is now virtually no restraint upon the use of rape products except the capacity of the market to consume them, but a recent professional estimate of the future of the crop in Europe suggests that E.E.C. support is likely to decline at the end of the 1980's (23). Mountains of butter, hillsides of beef, avalanches of grain and lakes of wine testify that there is need, in Europe, to reform the management of agricultural resources with an eye to social and economic advantage rather than mere expansion of output. By 1984 E.E.C. expenditure on rapeseed support had increased by 400 per cent at a cost probably equal to or greater than that of supporting growers of fruit and vegetables or grapes or olives. Amongst these latter there are, throughout Europe, many more disadvantaged producers than amongst the growers of rape and the social advantage of assisting the less fortunate is clear. Farmers doing well on rape probably also do well on creals or sugar beet and so it is reasonable to suppose that support might be transferred from them to growers of olives, fruit, or vegetables. If this were to happen the acreage put down to rape in the U.K. would presumably fall (23).

However, Europe is at present far from self-sufficient in either vegetable oil or oilseed meal. The demand for rapeseed products remains high and the prospect of widely available 'double-low' varieties improves the outlook for the European producer. Moreover, the United Kingdom has climatic advantages for rapeseed production beyond those of the more southerly areas in France and Italy which are better suited to produce sunflower or soya bean. Such a distribution of oilseed production is likely to ensure that rape production continues in Britain for the foreseeable future (6).

For beekeepers to be able now to rely, for an indefinite number of years to come, upon the regular annual availability, in May and June, of a huge new nectar source, is for them an historic advance. The pattern of work through the beekeeping year is changed by it and a new yearly flow of nectar and honey is in prospect with a yield comparable to that from the heather. That there are problems for the apiarist connected with the exploitation of rape is in keeping with the rest of his experience but the odds against him getting a profitable advantage are not greater than for any other honey crop.

Does rape need bees?

It is a curious circumstance that, although seed-bearing crops provide the staple of human food almost everywhere in the world, few of them depend for their productivity upon pollination by insects. The temperate grain-crops, wheat, barley and oats (and in the tropics rice, millet and maize) are, in common with all other grasses, wind-pollinated. Had it not been so, farming practice as it is known today in Britain would have been suicidal because the technical methods of modern agriculture lead inexorably towards the extinction of wild, insect pollinators. Large, complex and expensive farm machines operate more and more efficiently and more cost-effectively as fields are made bigger and hedgerows grubbed out. Herbicides, insecticides and standardized drainage make possible huge areas of monoculture where only one plant-species survives: the crop. Other species are suppressed. It is a process of selective cultivation which results, amongst other things, in the destruction both of the nesting-places of wild, pollinating insects and of their natural, native food-plants, the attack upon wild plants and insects leaving farm productivity apparently untouched.

However, had wheat, barley and oats not been wind-pollinated plants but insect-pollinated instead, the story would have been quite different. To kill the pollinators or destroy their food would have meant famine.

The introduction of a permanent, widely-grown, yearly crop, highly attractive to insects might, at first sight, seem likely to re-stablish insect pollinators as a valuable natural resource, but it is uncertain to what extent rape crops benefit from the visits of pollinating insects. Experimental work in this connection, though extensive, is far from conclusive and there is little evidence, in England at any rate, that the rape seed crop is much improved as a result of cross-pollination by honeybees (21, 48). However, scarcely a farmer is to be found who does not believe that having bees in the rape is advantageous and there is also a number of experiment-based reports, mostly from outside Britain, that pollination by bees can improve seed yield (1, 30, 32). Aspects of the natural history of the plant also show how knotty the problem of rape pollination really is (18).

The seed is sown in late August or early September and, given suitable weather, germinates and grows throughout the autumn, winter and following spring. It matures as a branching, three-and-a-half foot, sea-green plant with toothed or lobed, succulent leaves. By August, when the crop is harvested, long, seed-carrying pods have replaced the numerous, bright yellow flowers.

Flowering begins in the second week of May and continues, dependent upon weather conditions, for some 25 or 30 days. The four-petalled bloom with its greenish calyx opens to expose the reproductive structures and hidden deep amongst them in the base of the flower are four bulging, green nectaries. High daytime temperatures tend to shorten the flowering period and unseasonal cold makes the flowers fall.

The complexity of the flower itself hints that its functions are not simple and to say that it provides for pollination and the production of seed is true but hides our ignorance of much to do with it.

The female reproductive structures, enclosing the plant-eggs or ovules, primed for the arrival of pollen and equipped to nourish and protect developing seeds, consist of a central two-chambered ovary surmounted by a slender pillar with a specialized pollen-receiving tip.

The male structures, encasing ripe pollen, specially constructed to burst open and spill out the grains, encircle the central pillar in two sets: an inner ring of four and another outer pair. The stalks of the inner group are long enough to raise the pollen-bearing heads to the level of the tip of the female pillar. The pollen from them is released towards the outside of the rim of the flower. The stalks of the outer pair are shorter so that the pollen-bearing heads nestle relatively low in the flower betwen the petals and the stalks of the taller neighbouring stamens (18). Pollen from these shorter stamens is released inwards in the direction of the centre of the flower and presumably is not as instantly accessible to agents of pollen dispersal (wind, insects, rain splash, natural move-

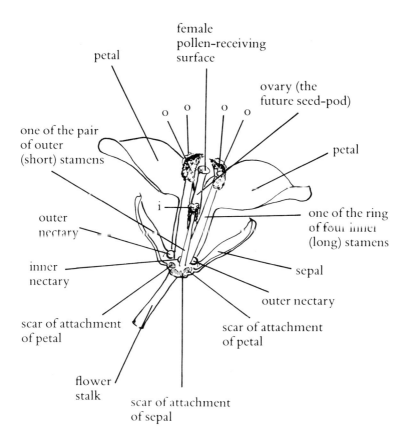

petal

female
pollen-receiving
surface

ovary (the
future seed-pod)

o   o   o   o

one of the pair
of outer
(short) stamens

petal

one of the ring
of four inner
(long) stamens

outer
nectary

i

inner
nectary

sepal

outer nectary

scar of attachment
of petal

scar of attachment
of petal

flower
stalk

scar of attachment
of sepal

Fig. 3. Oilseed rape flower with one sepal and two petals removed from the front. One sepal, one short stamen and one of the inner pair of nectaries are hidden at the back.
o = outwardly directed pollen   i = inwardly directed pollen

ments of the flower itself) as is the pollen from the four long stamens.

At any rate, in any single individual plant, pollen has two different origins: the long stamens and the short (18). Each kind is capable of reaching the stamens of its own flowers and bringing about self-pollination or reaching the flowers of other individual rape plants to cross-pollinate them. This means that in any field of flowering rape one can distinguish four modes of pollination:

13

1. Cross-pollination by pollen from the short stamens
2. Self-pollination by pollen from the long stamens
3. Cross-pollination by pollen from the long stamens
4. Self-pollination by pollen from the short stamens.

Work has been carried out, though not in Britain, which examines the effect of each of these different modes upon the weight of seed per pod. The above list aranges the possibilities in what proved to be the descending order of productivity. Best results were obtained when short-stamen pollen was used in cross-fertilization. The weight of seed per pod was then 19% greater than when similar pollen was used in selfing. Long-stamen pollen gave lower seed weights by 14% when used in crosses, and by 12% when used in selfing (18, 39). Were it possible to arrange for only short stamen pollen to be available in the field the benefit would probably be marked but, since long-stamen pollen is necessarily there also, any advantage is largely lost since the chances are against the most beneficial mode of pollination occurring in any particular flower.

That cross-pollination does in fact occur has been shown by using rape varieties with particular characteristics (e.g. unusual petal colour) which cross-breeding with normal yellow-flowerd plants causes to disappear in the following generation. These varieties are grown surrounded by plants of normal-coloured rape so as to ensure that cross-pollination can readily take place. Rape seed from the test-variety is collected and sown in the following season so that the colour of the flowers of the new generation can be discovered. The more the unusual petal-colour has disappeared, the greater was the amount of cross-pollination. Such experiments showed that between 20% and 42% of the test plants were affected. However, the number of honeybees flying in such experimental plots seemed not to change the outcome. Even if the bee population was increased five-fold no important resultant difference in the amount of cross-pollination was found (39). It is difficult not to conclude that, for the rape varieties used in these and other trials, insect pollination was of small importance and that the main agent of pollen transfer was the wind (31, 50, 53). Not all varieties of oilseed rape are alike in this respect however. In some the pollen grains are

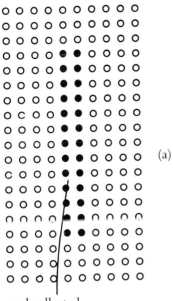

(a)

seed collected

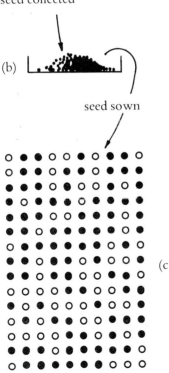

(b)

seed sown

(c

Fig. 4. Representation of an experiment to estimate the amount of cross-pollination taking place in a plot of oilseed rape.

(a) A plot of rape in the centre of which are two rows of plants with flowers of unusual colour. The unusual petal-colour is a weak genetic feature which is overcome by the stronger 'normal' colour when cross-pollination takes place between the two varieties, but which survives if self-pollination occurs. The plot is exposed to insect pollinators and seed collected from ripe fruits from the two centre rows.

(b) Collected seed.

(c) Plot sown, during the next growing-season, using only seed collected from the two centre rows in (a). All plants with flowers of 'normal' colour are the result of cross-pollination. In this imaginary case 33-34% of the seeds were produced after cross-pollination.

15

reported to be sticky and unlikely to be wind-transferred but to be flicked about from the springy anthers when conditions are dry (14).

Cross-pollination, however, is not merely a matter of getting a pollen-grain from the stamens of a flower to the stigma of the flower of another plant. There are many complicating factors such as the genetic variety of the rape, the age of the plants, the state of the flowers, the weather conditions both generally during the season and also at the particular time of transfer when the pollen is most exposed, the viability of the pollen and the rate of its growth after arrival.

Such questions seem not yet to have been much explored in connection with the pollination of rape but however incomplete such information might be, there is no doubt of the attractiveness of the crop to honeybees. Indeed, the question arises how such devices as exist in the rape flower for the attraction of an insect pollinator can, apparently, at once be so effective and so functionless.

A pollen-grain sends out a long tube which must grow through the female tissues from the stigma to the ovary of the flower it is to fertilize. Several pollen grains may do this at once and, since they differ genetically in their rate of pollen-tube growth, vie with one another in a sort of competition or race where the winner successfully fertilizes the ovule.

Forage

A good deal of attention has been paid to the behaviour of bees in flowering rape and it is well known that a number of species energetically exploit the crop. Honeybees, showing the economy of effort usual in nature, concentrate their numbers in the vicinity of their hives so that whilst four or five thousand per acre may be working in the rape within two miles of home, the density of the foraging population falls steadily as distance from the hive increases (2). Such economy is readily understood since the same sugars are used as flying-fuel as are collected as nectar and brought home to store so there comes a point when it is wasteful to fly further since more food is used in making the journey than is collected. But it is hard to imagine that a rape crop has ever been fully exploited by bees. One estimate puts the number of flowers at about 4 million per acre (44). Even when 90% of flowers have shed their petals foraging bees still persist in their operations in undiminished numbers (19).

Systematic observation of the bees at work in rape has shown that only a small minority (up to about 5%) concentrate exclusively on pollen-gathering. Only occasionally were the purposive, scrabbling, pollen-collecting movements seen although, of course, many foragers by chance became dusted over with grains as they moved in the flowers. Some individuals regularly packed their corbiculae with pollen whilst others discarded it. Foraging began at about 6.30 a.m. and, though most pollen was collected between 8 and 9 a.m., continued, in good weather for some 12 or 13 hours. Though rather fewer bees visited the crop when flowering was advanced, the proportion carrying pollen loads increased, perhaps because towards the end of the flowering less nectar is available (21).

However, the primary objective of bees foraging in the rape is to reach the floral nectaries. These are four solid, bulging glands of green tissue in the base of the flower and from the surface cells of each oozes the sugary liquid sought by the

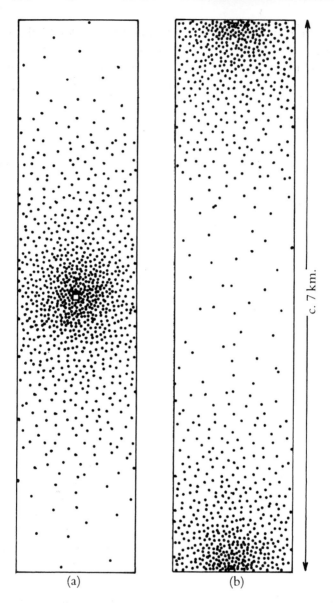

c. 7 km.

(a)   (b)

Fig. 5. The distribution, in good weather, of foraging bees rang-
ing over flat land where a rich nectar source is evenly spread. The
dots indicate the concentration of foragers from place to place and
show how this varies with distance from the hives (a) with a single
central bee-community and (b) with two communities separated by
about 7 km. (4 miles).

Based on figures from Benedek, 1976.

18

bees. The process of secretion is relatively slow and about 30 minutes is required for a nectary (32) newly drained of its product, to become flooded once more with a fresh drop of nectar. The nectar commonly contains 30-40% of sugars though there is variation and as much as 60% is recorded (14). The factors governing the production of nectar are numerous and their interplay complex. The characteristics of soil and climate, length and time of day, duration of sunshine and even the frequency with which the nectary is drained by the bees all have an effect. The latter is of practical interest to the beekeeper since the oftener a nectary is emptied by the bees in the course of a day the more nectar it secretes (18).

But not less important than any of these is the genetic variety of the rape. A number of tests show that the yield of nectar varies from type to type and much interest must now attach to the new 'double-low' varieties available in England. 'Darmor', marketed by the French in 1984 and 'Mikado', an English-bred variety appearing in 1985, are almost certain to become more widely grown than at present but are yet to be proved as sources of nectar and honey.

The productivity of the nectaries is unequal. The two innermost, which lie at the bases of the short stamens, produce much more nectar than the outer pair and there may also be differences in the mix of sugars or in their concentration. At any rate the bees distinguish sharply between them and show the strongest preference for attending the inner pair (14). One observer examined 50 rape flowers on a fine day when the bees were flying and found 41 inner nectaries, but only 17 outer ones, empty. Also, of 50 bees seen to alight on a flower only one failed to explore the inner nectaries, whether they were full or not. Half these bees, finding the first nectary empty, flew off without inspecting the others (32). Other records show that when flowers of rape with all nectaries filled were exposed to foragers, only nectar from the inner glands was taken. Nectar from the outer pairs remained untouched. The filling of nectaries can be brought about by isolating flowers from the bees for an hour or two (32). This preference on the part of foragers is striking, particularly since the outer nectaries appear to be at least as easily reached by the bee proboscis as the inner and

19

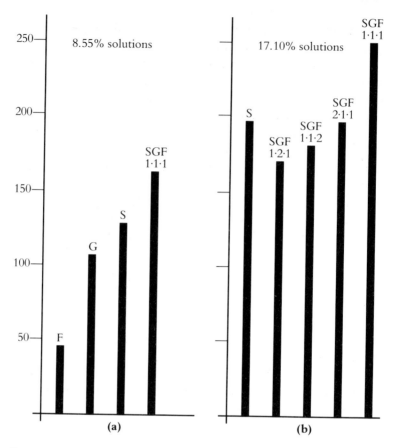

Fig. 6. That bees distinguish clearly between sugar solutions of different composition is shown in these results from two experiments to count numbers of visits paid during a measured period if time by foraging bees to pools of various solutions of single sugars or of sugar mixtures. S=sucrose, G=glucose, F=Fructose. Numbers attached to the letters indicate the proportions of different sugars in the mixtures. Concentration of sugar solutions (a) 8.55%, (b) 17.10%. (From figures given in Wykes G. R., 1952.)

In each case bees favoured solutions containing equal quantities of the three sugars.

may even be specially accessible when the flower is fully blown.

Foraging for nectar is usually, but not necessarily, accompanied by pollination. A proportion of bees (17-24%) have been observed to make a way into flowers, by a side entrance as it were, pushing their tongues between the bases of the floral parts. No contact is made with the receptive female surface in the centre of the face of the flower and no pollination by contact occurs (20).

In taking rape nectar the bees are collecting a plant-product with distinctive characteristics. All nectars are complex mixtures of substances dissolved in water. They originate in and are derived from the sap of the parent plant which is processed, changed and secreted by the glandular cells of the nectaries. The predominant components, apart from water, are the three sugars: glucose, fructose and their combination product, sucrose.

Glucose and fructose are examples of the group of simplest sugars (the so called 'monosaccharides'). The former is found in the blood of all animals and in the sap of plants; the latter in fruits and in other plant structures. Sucrose is the ordinary cane-sugar of everyday domestic use. They are the substances upon which almost all living things rely as sources of life-maintaining energy, warmth and action.

The relative proportions of these three, combining together in any particular nectar, vary between one plant species and another. In some, mainly (or only) sucrose may be present. In others there is a combination of all three. Oilseed rape nectar however, contains predominately glucose and fructose. Sucrose and some other sugars are present only in very small quantities (10).

The amounts and proportions of these sugars in rape nectar and, subsequently, in the honey from it, entail a legacy for the beekeeper. They determine the nature of the honey he is to harvest and oblige him, if he is to harvest it successfully, to adapt his programme and his methods to its uncompromising properties.

SECTION IV
Crystallization

Between its arrival in the hive and the closure of the honey cells by the bees, nectar, whatever its plant of origin, is 'ripened' into honey. An important feature of the ripening process is the progressive expulsion of water from the nectar, the main cause of the change being the activity of the bees in maintaining a high hive temperature by their own physiologically generated heat and in circulating currents of fresh, relatively dry air through the internal spaces of the hive. This, during the course of one or two days, progressively concentrates the liquid so that the proportion of sugars in it steadily rises. The change represents an important domestic economy for the bees because the more concentrated the liquid stores can be made, the more life-sustaining nourishment can be packed into the space available in the combs. In addition, during the collecting and ripening process, the bees themselves produce and mix with the nectar small quantities of chemically active substances (enzymes) which have a special power to transform sucrose into a mixture of the two simpler sugars, glucose and fructose. This change, sometimes called 'inversion' of sucrose, means that the resultant product is richer in energy per unit volume because the total water-solubility of fructose and glucose together is greater than that of the sucrose from which they are produced. A more concentrated honey can be made and its food-value increased in proportion to its bulk.

Inherent, however, in the changes in mix and concentration of sugars which the bees bring about whilst ripening their honey is a risk; namely, that of crystallization. Of course, honey is made for bees, not for beekeepers, and it is hardly surprising that the risks inherent in granulation in the hive are greater for intruding man that for the honeymakers themselves. How the making and use of honey by insects began is hidden from us in the remote distances of evolutionary history but has been going on so long that the adaptation or 'fit' between the behaviour and responses of the bees, on the one hand, and the peculiarities of their purpose-made

22

food on the other, must be almost perfect. Perhaps too little attention has been paid to the behaviour of bees as an interpretation of the nature and characteristics of the honey upon which they so closely depend.

In the ordinary way crystallization does not occur within the hive because the temperature is either too high or, perhaps surprisingly, too low; or because stores are used up before their condition can change. In stores of oilseed rape honey, however, the risk of rapid, hard crystallization is so high as to be almost a certainty and, if they are left in the supers for longer than about ten days after capping, are unlikely to be got out, by man at any rate, without destroying the comb. Experienced beekeepers who regularly exploit the rape crop say there is no important immediate threat to the bees.

Observations made upon honeys of average composition show that crystallization occurs in them most rapidly at 14° or 15°C (12). Either side of that temperature the tendency to granulate declines. Above it, if they appear at all, crystals tend to be coarse and gritty whilst at temperatures increasingly further below 14°C the speed of crystallization becomes less and less. At 10°C it is retarded, at 4½°C is almost stopped and at −1°C no sign of crystallization is seen even after 2 years (12). Every beekeeper is aware of the crystal-banishing properties of heat but it is, at first sight, perhaps unexpected that cold also prevents granulation. However, the colder liquid honey is made the stiffer, thicker and less mobile it becomes. A state is reached when, not only is the flowing motion of the honey becoming extremely slow and viscid but, simultaneously, the activity needed to build up crystals, the minute, internal dancing and diffusion of chemical particles, is being depressed. Under such conditions no crystals can be built up. But if honey is already granulated at, say, the favourable temperature of 14°C, no amount of cooling thereafter will reduce the granulation but, on the contrary, will consolidate it.

It is very striking to notice that the bees themselves also react strongly at 14° or 15°C. At about that temperature their behaviour undergoes the marked changes leading to the formation of compact winter clusters. The adoption of cluster formation is reported not to be sudden but to begin when temperatures fall below 18°c and to be complete only at 0°C to −5°c (40).

One result of clustering is, presumably, the exposure to the temperature of the atmosphere of all combs in the hive except those parts covered and protected by the bees. All physiological heat originating with the bees and previously affecting all parts of the hive where they were present must be dispersed except within the cluster and perhaps sometimes also in areas immediately above the cluster which might be warmed by convecting air. If atmospheric temperature is, at the time, falling, as tends to happen with the approach of winter and which is anyway the cause of the clustering, stored honey may not for long remain subject to a temperature of 14°C, but rapidly cool below it. On the other hand, any honey within the cluster will have its temperature determined by the bees and they are well known to maintain a temperature of 30°C, even when broodless. At 30°C no honeys granulate (47). It is reasonable to suppose that in this way the crucial, optimal crystallizing temperature of 14°C is largely avoided throughout the hive so that the honey tends to be kept in its liquid state.

Even if this speculation is true and even if there is such a honey-protecting adaptation in the honey-bee, it fails to prevent granulation in the hive of honey stores from oilseed rape and certain other plants.

There is, of course, no doubt that bees can exploit solidified food-sugar. The routine use of candy shows this to be true. There is also evidence (24, 25) that even dry, lumpy sugar can be transformed by bees into liquid stores particularly in times of dearth. The process is carried out only by nurse bees but, if they are to do it efficiently, they require a supplementary source of moisture, either in the form of plain water or nectar from their foraging companions, or in the form of syrup being fed to them at the same time as the dry sugar. Moisture is pumped out through the proboscis onto the sugar surfaces, dissolving some of it. The lump can become pitted with holes where nurses have been repeatedly at work. The new syrup is treated like any other part of the communal reserve and is distributed throughout the hive. It appears as food in brood cells or as stores in the honey cells.

1. Oilseed rape at arms length. (Ted Hooper)

2. Rape honey is as white as milk. (Paul Taylor)

3. Never put a hive in the middle of the rape! (Ted Hooper)

4. The floral 'works' of Brassica napus. (Ted Hooper)

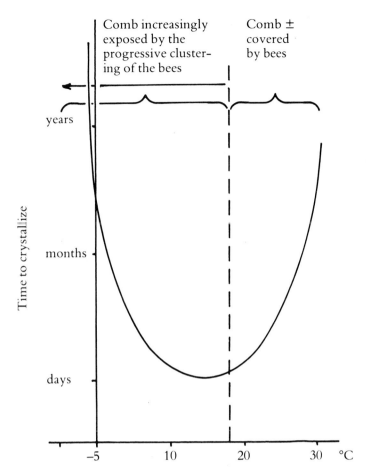

Comb increasingly exposed by the progressive cluster-ing of the bees

Comb ± covered by bees

years

months

days

Time to crystallize

−5 10 20 30 °C

Fig. 7. Illustration of the tendency, at different temperatures, of honeys of average composition to crystallize. The effect of temper-atures most favourable for rapid crystallization is perhaps largely avoided by the clustering of the bees at 18°C and below. Comb is then exposed to winter atmospheric temperatures which tend to be well below 15°C.

The transferences are shown up by staining the dry sugar with a harmless blue dye which then re-appears wherever in the hive the blue syrup is put down. The process can con-tinue so long as a sufficient moisture supply is maintained but, should the latter come to an end, dry sugar in the hive becomes, at best, functionless or, at worst, a burden since it can then only be dissolved by making inroads into 'physiological' water in the tissues and body-spaces of the hive bees – an expenditure the community may not be able to accommodate.

However, the use of crystallized or solid stores cannot be without its cost. No operation is carried out by the bees without the expenditure of energy and more actions are involved in making solid stores available as food than are needed when there is liquid honey to be had. A hungry bee, in taking nourishment from a crystalline source must carry out additional muscular operations, first to suck up water and then to expel it again onto the sugar surface. After that the ordinary procedures of ripening and distribution will follow. In addition, if crystalline food sources are being used, foragers are engaged in bringing in the indispensable water; an extra activity which, though essential, has no food or energy pay-off for the community.

From the beekeepers viewpoint the crystallization characteristics of honey are of great importance. They determine the procedures which it is practicable to follow in harvesting and bottling the product and have a strong influence upon sales. An understanding of them would be of great value but, though a good deal is known, it is far from complete.

Honeys are mixtures of very concentrated solutions of various sugars in water, with many other organic materials present in small amounts. The physical chemistry and behaviour of such systems is complex and obscure because the numerous components interact with one another, each responding to the environment provided by the presence and characteristics of the rest. However, the experimental work of honey scientists and physical chemists has allowed some general rules to emerge which indicate what are the probabilities of crystallization.

The most reliable general indicator of the tendency to granulate is the 'dextrose/water ratio' (12). This determines that, other things being equal, the more dextrose a honey contains, the more likely it is to crystallize. (N.B: dextrose and glucose are the same.) In oilseed rape honey the proportion of dextrose is unusually high (10, 12) and on this ground alone rapid granulation is to be expected. But rape honey also contains a high concentration of fructose. (N.B: fructose and and laevulose are the same.) And this has the effect of further reducing the solubility of glucose and, consequently, increasing the likelihood of crystallization. The influence of the concentrated fructose is thought to increase suddenly if temperature falls to 30°C or less (26).

It is worth noting that 30°C is also the normal temperature within a covering layer of active bees or within a winter cluster. Presumably the departure of bees from any part of a comb filled with rape honey causes a sudden, special increase in the likelihood of it crystallizing.

From the apiarist's viewpoint

Whatever the state of our understanding of oilseed rape or of its honey there can, nowadays, be no escape for many beekeepers from the regular annual presence and influence of the crop. In many areas apiarists are obliged, willy-nilly, to take into account a yearly input of rape nectar into the hives simply because the rape is all about them. Modern, large-scale agriculture has often so changed the face of the countryside that no other major nectar source is left. Under such circumstances there is no alternative but to adapt to the new conditions if beekeeping is to continue as a serious pursuit.

Large-scale farming and the annual planting of fields of rape have confronted beekeepers with a changed floral year. The old, long-standing reliance upon the regular annual succession of sycamore, hawthorn, clover, bramble, willow-herb and heather has gone and, where the rape crop now plays a major part in farming, there is often very little bee forage between it and the heather. Hedgerow, roadside, coppice and wasteland flowers may largely be absent. As a result many large-scale apiarists have reorganized their yearly sequence of operations to take advantage of the two major resources, rape and

Good hive-sites in the rape will be

1) *free from the threat of theft or vandalism*

2) *remote from public roadways*

3) *as close to the oilseed rape crop as possible*

4) *as well sheltered as possible*

5) *situated so as to allow access for transport.*

28

heather. Without this reorganization, which is just as necessary for the small operator as for the big, the flow from rape is at best a nuisance and at worst a disaster. The reorgnization has to take into account:

(a) the regular arrival of an early, high-volume honey crop;

(b) the uncompromising properties of the honey, demanding either quick extraction at an unusual and perhaps inconvenient time of year or a modification of extraction methods if there is to be delay in doing the work;

(c) the effects of rape honey residues in the hive.

New crops require beekeepers to find new apiary sites and good places are difficult to find in the rape. Experience seems to show that, of the numerous desirable site features, shelter and freedom from thieves and vandals are the most important. But hives should not be put down in the shelter of hedges verging public roads. Any apparent isolation is misleading and passing people or horses may be stung with resultant angry complaints and damage to public relations. Besides, hives in such places are often spotted by miscreants and losses due to theft and vandalization have much increased in the last ten years. Uncultivated waste land,

Well-founded hives in the rape will

1) *be dry. If the place is damp, hive-stands or raisers will be used*

2) *be weed free with the front ground-areas covered by a weed-smothering board, metal sheet or polypropylene sack*

3) *not be smothered or overgrown by rape plants, but 'in the clear'*

though not quite vanished, is difficult to find but vacated railway-lines have, at times, provided the ideal conditions of shelter, dryness, relative remoteness and even a little subsidiary forage for the bees as well. It is useful in this connection, to notice that rape will not grow in wet or waterlogged places and, occasionally, bare or very thin patches can be found in the fields where the soil is too wet for rape to thrive. If these are near a non-public road they can be exploited, but warnings are given never to place hives so that the growing crop surrounds them closely since the plants rapidly grow so tall and so tangled together that hives can later be got out only with greatest difficulty or by scything away the rape from around them; an operation unlikely to win the farmers approval. It may be necessary to put down battens or hive-stands to prevent the hives getting damp.

In the shelter of a growing rape crop agricultural weeds often proliferate and grow luxuriantly. To prevent hives becoming overgrown it may be useful to put them down on some weed-smothering base. Galvanized sheets, 2 feet by 2 feet, have been successfully used to provide a bare space or, alternatively, the hive front area may be covered with a jute or polypropylene sack to keep down the weeds.

Consultation with farmers and farm managers about sites is advan-

Good management in the rape should seek to provide

1) slow stimulative feeding in fore-going August

2) queens not older than 'second season'

3) maximally strong stocks

4) frames of good drawn comb

5) queen excluder and super(s) in place from start of flowering (or from placement of hives)

6) bait hives, with good drawn comb, put out at rate of 1 for every 4 working stocks

tageous as well as politic. It saves time and effort and is good for public relations. There is general approval of having bees in the rape and the belief is widespread that the crop benefits. The fruiting of the rape is thought to be more uniform when the bees are there. Some believe that it is increased. There is an impression, however, of some relationship, not yet properly understood, between the density of the crop and the yield of the honey from it. Sometimes bees will not forage intensively in very dense stands of rape whilst, on the other hand, the most unimpressive fields can be the heaviest croppers for the beekeeper.

The nectar flow from oilseed rape arrives when stocks are emerging from their winter state and, though some strains of bees respond better than others to the early flow, good management in the previous season is important. Queens should be young and, wherever possible, replaced annually. Slow stimulative feeding of stocks in the foregoing August is better than feeding in the spring and if the location (and the weather) is favourable reproductive activity can be started up by the use of pollen substitutes. Such preparation allows the beekeeper to take maximum advantage of an extremely valuable, additional early asset which, in spite of its inherent drawbacks, is a source of volumes of honey, of large stocks of bees and of good supplies of drawn comb and wax.

If the apiarist's interests are mainly in the production of bees for sale, or in the production of drawn comb, he will probably prepare the maximum number of prospectively strong stocks in the second half of April by uniting and 'equalizing' any that are weak or below par. These are procedures that are in the ordinary way of good beekeeping and beekeepers with only a few hives are best advised to take the trouble to do the same so as to give each stock destined for the rape a strong start. But there will be many spare-time operators who, because of the little time at their disposal, have no alternative but to take their hives to the rape just as they happen to come through the winter, put on the supers and let them take their chance.

The importance and value of drawn comb is sometimes not appreciated by beginners. Experienced keepers sometimes declare it to be 'the next best thing to bees' and of particular value in heather honey production. The reason for its importance is that it is derived from the same raw material as the honey itself, namely, from the sugars of the incoming nectar. If a part of the nectar input of a colony has to be diverted for the manufacture of wax, as must happen if a hive is short of the wax-walled storage cells, then so much less is available for brood rearing and honey production. If, on the other hand, drawn comb is present throughout the hive, sugar input can be devoted entirely to building up brood or to the accumulation of honey.

Of all types of honey, that from heather has the highest market value, so any means of increasing yields from the heather brings extra profit for the beekeeper. A good season on the rape and successful extraction of the rape honey leaves him extensively equipped with frames of drawn comb and ensures that the honey yield he gets from the heather will not be depleted because his bees have had to make a lot of wax comb in which to store it.

In addition, the presence of good, empty, drawn comb in the hive gives the bees the extra security of a favourable supporting surface upon which they can get a good purchase during the disturbance of a move, and some experienced operators also consider that it reduces the tendency of stocks to swarm.

Fine weather, flowing nectar and abundant pollen mean increasing bee stocks in the rape just as in any other rich floral growth. But in rape the stocks build up two or three weeks earlier than in British countryside where rape is absent. The same beekeeping responses are called for but at an earlier date than operators are ordinarily used to. Whether the appropriate work can be carried out, however, depends upon the scale of the beekeeping and the time and labour at the beekeepers disposal.

Bee farmers with several hundred hives and a living to earn by their work are obliged to keep down daily costs to a minimum and to reduce labour-intensive operations as far as ever possible. But the spare-time apiarist may be just as constrained to make economies in his work and may find it to his advantage to take leaves out of their book.

5. Spraying is in fashion when the rape is out of bloom
(Michael Singleton)

6. Hives on the headland (Paul Taylor)

7. Good access is important. (Paul Taylor)

8. Golden acres. (Paul Taylor)

So, for example, when stocks are first put down in the rape, queen excluders and supers of drawn comb can be put on them there and then. Not only are future visits for the purpose of supering avoided but, in the view of some, the tendency to swarm is reduced. If conditions are good the build-up is so rapid that queen cells can appear in a matter of days. Supering, perhaps with two or even three supers over a strong stock with four combs of brood or more, can delay the construction of queen cells.

Or complication may be lessened by using deep boxes only. This can be done where the objective is to get a large brood area and lots of bees and where honey is a secondary matter. For this purpose, the first box is allowed to build up until about seven combs are covered by the bees. Then a second deep can be added without excluder. If to get honey is the objective then a queen excluder is put in. Here, of course, the work cannot all be done during a single initial visit.

Some large-scale beekeepers are so busy in June, perhaps with the marketing of honey for the onset of the tourist season, that nearly all manipulation of hives has to be abandoned and swarming just allowed to go ahead. A common counter-measure is to put out bait-hives, i.e. complete, unoccupied deep boxes with wide entrance and with frames of drawn comb. They are put out at the rate of one for every four or five working stocks. If the comb is in good condition from the previous year, and not decrepit and ruinous, the chance of swarms entering is quite high. Small beekeepers with 'time problems' should consider using them.

Where a good number of hives has been put out the observation is sometimes made that swarming is localized in a particular group, as though the first swarming stock had transmitted the impulse to its immediate neighbours. Beekeepers who find this give extra attention to hives in the vicinity of a swarmer.

The management of apiaries during full flow from the rape depends mainly on the scale of the operation. For the bee farmer with perhaps a thousand hives it is usually out of the question to carry out much in the way of inspection, manipulation or swarm-control. Sometimes visits are paid when, after putting aside the supers, the hives are tilted so as to expose to view the bottom bars of the brood chamber frames. The absence of queen cells is taken to indicate a stable stock not set on swarming. Often, rather than carry out more manipulation in stocks when swarming seems likely, bait-hives are put out. These are frequently found occupied in mid-June but, inevitably, stocks are lost from time to time.

However, a lively stock coming to its peak in the rape in late June presents problems if the bees are not to be sold. If it has been stripped of honey and if, because of local farming practice, there is to be a dearth of forage until the heather flows in August, the bees may have to be fed. Also, if the queen is not in her first or second season she may be too spent to sustain a strong population in the latter part of the year and may need to be replaced.

Where the main objective of the year's work is to get the maximum crop of heather honey, the rape may be regarded mainly as the means of building up colony strength in the spring and only secondarily as a source of blending honey for later use. The very earliness of the nectar flow from rape means that many bee stocks, when they receive it, are in a depleted condition after the losses of winter and will have increase in numbers as their first priority. The abundance of pollen from the rape also ensures that increase is rapid and that the energies of the hive are firstly directed to population growth. On the other hand, the last fortnight of flowering is the most important for the accumulation of a surplus of rape honey.

Before the rape crop itself is cut and harvested the bees must be moved and this is heavy work in a good season. A broodbox with two or perhaps three full supers cannot all be lifted at once. Provision needs to be made for clearing the supers first. This can be done, for example, by putting in an empty super, with crown board and Porter escape, between the broodbox with its queen excluder and the full supers. In 36 hours or so the supers, cleared of bees, can be removed and the broodbox returned to winter quarters.

Beginners should notice that moving hives from place to place is not without its hazards for the bees, particularly if the journeys are to be long. Bees should not be moved more than short distances in empty boxes or in boxes fitted only with frames of undrawn foundation. Long journeys are best made with the bees on drawn comb where they find maximum support and security. Recommendations are made to move stocks early in the morning or at night-time so as to keep temperatures down as far as possible and minimize the penetration of light into the stocks. Screenboards need not be used, but if crownboards are exposed, the Porter escape holes should be covered to exclude light during the move.

Much of the nectar obtained by the bees from the rape is, of course, at once consumed. Surplus is stored as honey and may crystallize. The higher the temperature within the hive the longer delayed will be the onset of crystallization and it is said that rape honey in a powerful colony will stay liquid for as long as six weeks. But if the colony swarms, suddenly reducing both the number of bees in the hive and also the hive temperature, any honey left behind will crystallize at once. This summer-time crystallization is not an irretrievable disaster for the bees because, provided that water supplies are available and flight possible, the crystalline stores can be used. In winter-time, however, crystallized stores of rape honey are largely useless. They are extremely hard when set and, because the bees are not flying freely to bring in the necessary water, crystalline material is simply left alone or thrown out.

From the beekeepers' viewpoint coping with rape honey is probably most affected by his having (or not having) access to a thermostatic warm space. Such a facility allows him various advantages. He can choose when he allows crystallization to occur because he can control the temperature of his honey. He can give any unripe stores time to mature and thicken in the comb, and he can soften or re-liquefy, at a time to suit himself, any stocks of granulated honey he may have from earlier harvests. In the absence of such a warm room or warm cupboard the rape beekeeper is obliged, in June in each year, quickly to extract and process, perhaps to completion, any rape honey he harvests. The coincidence of this inescapable chore with the onset of swarming and the maturation of new queens is the crux of the problem of oilseed rape. Either the extra work has to be done at the time or else changes in extraction procedures must be made so that the honey can be got later in the year.

Action begins with the collection of combs from the supers. In a good season the input of honey into the supers is rapid and visits to remove comb may be paid every two or three days. Because rapid accumulation of stores and urgency to extract the honey coincide there is a strong risk, as all operators agree, of harvesting unripe material and the subsequent unfortunate likelihood of fermentation in tin or jar and of spoiling the flavour of large batches of honey. Honey is unripe if it drips from the cells when comb is taken out of the hive and held horizontally for a moment. Such comb should be replaced amongst the bees. Harvested combs which are not fully capped, but do not drip thin honey, are routinely treated in the warm cabinet, where one is available, to a temperature of 95°F (35°C) for three or four days. Anything more than half sealed is then fit to extract.

If rape honey is to be extracted successfully without sacrifice of comb there is no doubt that the work must be done without delay. Estimates vary somewhat, but 10 to 14 days is commonly said to be the maximum interval of delay once rape comb is removed from the hive. It is interesting to notice, however, that new wax appears to have a restraining effect upon granulation. One large-scale bee farmer finds that rape honey does not set in virgin comb for a month and there is direct experimental evidence to the same effect (17). It follows from this that, if there is pressure of work, honey in new comb might well be left until last to be extracted whilst any in drawn comb from earlier seasons should be got out of the way as quickly as possible. Ideally, extraction should be carried out on the day the comb is collected. The empty comb can then be returned to the hives for refilling. But whether on the large-scale or the small-scale the production of rape honey can be attended by product-losses not experienced with other honeys. The tendency to crystallize is strong and persistent. If, after use, equipment (including screens and straining cloths) is not scrupulously cleansed of films of honey, it will become clogged and unusable.

A film of honey remaining on the wax of extracted combs will crystallize and can bring about granulation in heather honey too if the same combs are used for both harvests. This crystal 'seeding' effect is very persistent and, if warming facilities are not available, the processing of heather honey should not be long delayed. In order to avoid such contamination of heather honey with rape crystals it is possible to reserve a set of supers and frames especially for the rape harvest though the cost probably makes it impracticable except in small apiaries. Even this precaution, however, may not succeed and there is an authoritative caution ary tale (9) of an experiment carried out in Scotland in 1984 where 20 stocks were sited 1.4 miles from a newly-established rape field. Supering was delayed until the rape flowering was over but, of course, rape honey would be stored in the brood chamber. Supers were put on in June to catch the mid-summer nectar sources and these yielded quite well, each stock finishing up with one or two well-filled supers. The crop was harvested at the end of September. (It will be noted that any rape honey still present in the stock had been there for more than three months.) Every super was crystallized to a greater or lesser extent varying from total granulation to 10% or so. Of a full total super weight of 668 lbs., only 142 lbs. of honey was recovered. The interpretation is proposed that, during the expansion of the brood nest, the bees had transferred previously-stored rape honey from brood chamber to supers where it had induced the granulation of the later crop.

As an efficient aid in the extraction of heather honey, whether or not following a crop of rape in the same combs, a form of loosener is recommended and regularly used by some apiarists. The principle of the device is the disturbance or stirring up of the heather honey in its wax cell by plunging once into it and withdrawing a thin metal rod of such a length as almost to reach the foundation of the comb. This reduces the viscosity of the honey sufficiently to ensure it coming out in the extractor. The simplest form is a hand tool with metal rods set in a wooden block. It is laborious to use but a mechanical version (the 'Norwegian loosener' or 'Norwegian needles') is obtainable which deals with each frame in a matter of seconds.

One most effective, wholesale solution to the problem set by the rapid crystallization of oilseed rape honey is to let the crystallization take its course in unwired supers and leave the whole sequence of processing until winter when there are no other beekeeping preoccupations. Everything that the bees produce is sold and the operator has income from the wax as well as from the honey. Since the production by the bees of 1 lb. of wax costs them about 5 lbs. of honey, there is no immediate financial loss to the beekeeper provided the price he gets for his wax is at least five times greater per lb. than he gets for his honey.

In this system the operator is confronted, at the end of the year, with unwired frames of granulated honey set in the comb. This he cuts out and mashes down into suitable containers and, in the warm room, raises it to a temperature of 115°-118°F (47°-48°C). Comb, even if wired, collapses before this temperature is reached at about 105°F (41°C), but the reason for heating so much further is that, whilst crystals in other honeys tend to melt away at about 95°F (35°C), those of oilseed rape honey do not do so until 115°F is reached. The wax floats to the top of the warmed mass and is, in the commercial operation, separated from the honey in a suitably high-powered centrifugal machine (an industrial 'spin dryer'). The honey is then strained, creamed and put away. On the small-scale, the warmed material can be strained through screens or through cloth, though the risk of crystallization during cooling is still present and equipment must be washed immediately after use.

It is recommended that rape honey should be 'creamed' before putting away to await further processing. For this purpose it is necessary to have a suitable 'starter' (i.e. a reserve, from a previous season, of especially fine-grained honey). The huge number of tiny crystals contained in fine-grained 'starter', when dispersed uniformly through a batch of warm, liquid honey, usually at a temperature of 75°F (24°C), ensures that a correspondingly large number of centres of new crystallization is established and that the new crop of crystals will be small and very numerous. When cool and set the resultant mix will be almost as fine-grained as the 'starter' was and part can be kept for use in the same way in the next season. It is often said that the same starter or its derived product can be used continuously year after year, but this is not altogether true. When that is done crystal size

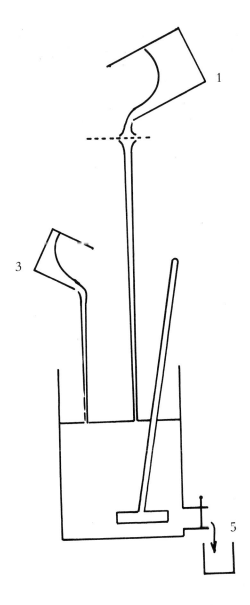

Fig. 8. Main steps in the 'creaming' of honey:
1. Warm bulk honey to 49°C and strain to remove wax and foreign particles
2. Cool bulk honey to 24°C or less
3. Add fine-grained 'starter' honey, also at 24°C and mix
4. Let stand to reduce air-bubble content
5. Run into containers

slowly but surely increases season by season so that the product gets gradually more coarse and gritty. To prevent this happening the starter is, from time to time, renewed from natural sources of fine-grained honey or else the gradually coarsening starter is ground in a suitable mill to reduce crystal size. The 'creaming' procedure, if effective, causes the formation in the new liquid batch of multitudes of tiny, flat, quickly-grown crystals of glucose. These tend to slide past one another giving the product a soft, smooth consistency. Big, slowly-grown glucose crystals are, by contrast, like microscopic hedgehogs and, when present, make honey rough, gritty and viscid.

Although it seems that no large-scale beekeeper seriously seeks to popularize pure rape honey, the market for it is not negligible. For many people its extreme whiteness, smoothness and bland, somewhat featureless flavour are attractive. It is produced by a rather romanticized figure – the local beekeeper – and if creamed and attractively bottled, has an appearance and flavour resembling some honeys from the supermarkets where the great majority of people do their shopping and acquire their ideas of what good eatables are like. Beekeepers with only a few hives should not neglect its possibilities.

Most honey from oilseed rape is used in blending with other honeys whose flavour is sufficiently strong to bear dilution. But blending, with its requirement for facilities for slow warming, seeding, mixing, containerization and storage, is often an over-complex and difficult operation except for the specialist. Even large-scale beekeepers sometimes prefer to sell their rape crop to a competent packer.

Bees and crop protection

During their foraging visits the bees may sometimes appear to have the rape to themselves but, of course, they do not. Co-partners with them in exploiting the crop as a food source are the various wild pollinators and also a number of insect pest species whose activities damage the plants and tend to reduce the yield of rape-seed to the farmer. The modern response is to attack these pests, when they are judged to be at their most vulnerable, with specialized insecticidal chemical substances sprayed onto the crop as mist-fine droplets, usually, in the case of oilseed rape, from light aircraft or helicopters.

The vetting of chemical substances proposed for marketing and use as insecticides is carried out under the Agricultural Chemicals Approval Scheme, a subsidiary organisation within the M.A.F.F. with access to the best scientific advice and with legal powers to give or withold permission for substances to be used by or on behalf of growers. As part of this scheme insecticidal materials are classified as 'dangerous' to bees (the most toxic and damaging category from the beekeepers viewpoint), as 'harmful' to bees, or thirdly, as 'presenting minimal hazard to bees when used as directed' (38). Before being sprayed onto crops from the air a chemical must also be specifically approved, by the same organization, for that particular purpose. Until 1985 revised lists were published annually of the substances in each of these groups (33 et seq).

Operating in close relationship with the Agricultural Chemicals Approval Scheme is another: the Pesticide Safety Precautions Scheme. These, between them, provide the public, including beekeepers, with safeguards against the irresponsible use of damaging agricultural chemicals.

As a result of the work of these organizations the M.A.F.F. is the main source of factual information about the effects of insecticidal sprays on bees. The A.D.A.S. National Bee Unit (Luddington) and the Tolworth Laboratories examine and analyse all reported cases about which evidence

41

(field information and dead bees) is available and which allege that poisoning gas occurred.

For example, in 1982 and 1983 together, they dealt with a total of 117 incidents, about half of which were from oilseed rape apiaries and affected an estimated total of 676 bee colonies (35, 37). In the great majority of cases triazophos was the implicated chemical. Probably these numerous disasters were due, mainly, to stands of oilseed rape being sprayed whilst still partially flowering and containing enough bloom still to attract the bees.

From another point of view, however, the M.A.F.F. seems increasingly (and somewhat surprisingly) to regard the organization of 'spray warning' as the beekeepers responsibility. It is interesting to compare the advice to farmers and growers published in 'Approved Substances for Farmers and Growers' from year to year. Thus:

1976, p.10: 'Get to know local beekeepers *and warn them* in advance of intention to spray and advise of any change in plans' (author's italics).

1980, p.8: 'Get to know local beekeepers *and their Spray Warning Officers (if appropriate). Warn them* well in advance of any intention to spray and advise them of any change of plans' (author's italics).

1983, p.9: 'Get to know your local beekeepers. *Ask them to nominate* a local spray liaison officer. *Tell him* in advance of your intention to spray and advise any change of plans' (author's italics). The emphasis thus shifts, in eight years, from the farmer to the beekeepers' organization where there is one (33 *et seq*)

Further, in the M.A.F.F. Consultative Document (November, 1985) from the Pesticides and Insecticides Control Division entitled 'Pesticides, Implementing Part III of the Food and Environment Protection Act 1985' the Ministry again appear to intend to rely on voluntary local organization to protect bees. Thus, under 'Regulation of Aerial Spraying' one reads '. . . Conditions are attached to the method of application which include . . . (c) the giving of advance warning to . . . local beekeepers *where a spray warning scheme is in operation*' (author's italics).

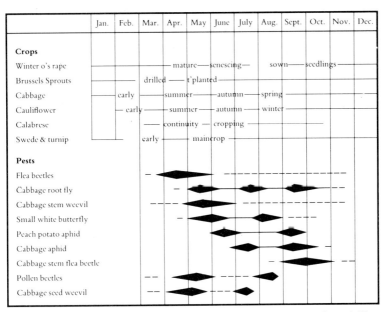

	Jan.	Feb.	Mar.	Apr.	May	June	July	Aug.	Sept.	Oct.	Nov.	Dec.
Crops												
Winter o's rape				— mature	—senescing—			sown	—seedlings	—		
Brussels Sprouts			drilled	— t'planted	—							
Cabbage	early			summer	—	autumn	—	spring				
Cauliflower	— early	—		summer	— autumn	—	winter					
Calabrese			— continuity	— cropping	—							
Swede & turnip		early	—	maincrop	—							
Pests												
Flea beetles												
Cabbage root fly												
Cabbage stem weevil												
Small white butterfly												
Peach potato aphid												
Cabbage aphid												
Cabbage stem flea beetle												
Pollen beetles												
Cabbage seed weevil												

Fig. 9. Synchrony of the main periods of brassica vulnerability and the times of pest attacks.

From Wheatley, G. A. and Finch, S., 1984. (By kind permission of the authors and BCPC Publications, Croydon.)

Most beekeepers would agree that the effective organization of a spray warning operation, with its need for mapping, communication, systematic daily intelligence and round-the-clock active implementation, is beyond the resources of the local association (where one exists) but that it remains a prerequisite so long as compounds dangerous or harmful to bees are in use.

Few local associations have much more at their disposal than a list of their members with addresses, phone numbers and the map references of apiaries. These they may put into the hands of cooperative spraying contractors. With good will the contractor does what is possible and practicable to protect the beekeepers' interests. Well-organized firms have technical sales representatives whose advice from the field determines when a particular spraying operation shall be carried out. They are in touch with the farmer-client, know the location of the crops, can estimate the state and vulnerability of the pest population from time to time (though rape is a difficult crop in this respect because of the density and impenetrability of the stands) and they have the responsibility of initiating spraying procedures when the effect is judged likely to be maximal. They, and senior staff to whom they report, have all the information most valuable to the beekeeper if he is to keep his bees out of the rape when chemicals are being applied. If the technical sales representative is also told the location and ownership of hives in the rape-crops in which he has an interest he becomes the ideal contact, amongst spray-contracting personnel, for apiarists.

The need for reliable spray warning, whether organized by the beekeepers or by others better able, is unlikely to diminish in the near future. The stabilization of rape acreage may well result in the enhancement and crowding-up of pest populations previously able to find new living-space in the ever-increasing volume of the crop. Pest organisms, leaving the senescent rape in June or July, could well invade other valuable brassica crops now closer to the out-reaching rape fields than ever before. Fig. 9 shows the relationship between peaks of insect pest activity and the life-cycles of brassicas they can invade (46). The threat they pose might best be countered by spraying in the rape.

However, beekeepers can afford some optimism about the future of crop-protection and its effects upon their bees. There are many signs that the message of environmentalists is being heeded. There is a tendency for insecticides to become increasingly selective, more perfectly targetted upon particular pest organisms and less hazardous to bees (42). Biologists continue to search for effective biological methods of pest control presaging, if not a farewell to chemical techniques, then a more comfortable partnership with them. Government departments spend more and more time and attention in seeking to improve the law protecting our natural resources, wild and cultivated.

Bibliography

1. Benedek, P. *et al* 1972. Insect pollination on oilseed rape. Növénytermetes, *21*, 255-269.

2. Benedek, P. 1976. Structure of flower-visiting insect populations and the effect of environmental conditions on honeybee density in blooming winter rape fields and their bearing on pest control. Növénytermetes, *25*, 65-74.

3. Bowland, J. P. *et al* 1965. Rapeseed meal for livestock and poultry. Canadian Department of Agriculture Publication 1257, 69-80.

4. Bowman, J. O. 1984. Commercial oilseed rape breeding. Aspects of Applied Biology 6, 31-36.

5. Bruckner, J. *et al* 1984. Rapeseed: constituents and protein products, Part 3: Nutritional value of protein enriched products. Nahrung *28*, 45-81.

6. Bunting, E. S. 1984. Oilseed rape in perspective. Aspects of Applied Biology 6, 11-21.

7. Butler, E. J. *et al* 1982. Problems which limit the use of rapeseed meal as a protein source in poultry diets. Journal of the Science of Food and Agriculture *33*, 866-875.

8. Cambridge Agricultural Publishing Company 1981. The oilseed rape book.

9. Couston, B. 1984. Rape and Spillage. Scottish Beekeeper, December, 210-212.

10. Crane, E. *et al* 1984. Directory of important world honey sources. International Bee Research Association.

11. Crane, E. 1975. Honey, a comprehensive survey. Heinemann.

12. Dyce, E. J. 1975. Producing finely granulated or creamed honey; in Crane, *op cit*.

13. Eddowes, M. 1976. Crop production in Europe. O.U.P.

14. Eiskowitch, D. 1981. Some aspects of the pollination of oilseed rape (B. napus L.) Journal of Agricultural Science, *96*, 321-326.

15. Fenwick, G. R. 1982. The assessment of a new protein source – rapeseed. Proceedings of the Nutrition Society, 41, 277-288.

16. Firth, C. H. and Rait, R. S. 1911. Acts and Ordinances of the Interregnum, 1642-1660. H.M.S.O.

17. Fordy, G. 1982. Private communication. University of Salford.

18. Free, J. B. 1970. Insect pollination of crops. Academic Press.

19. Free, J. B. and Ferguson, A. W. 1980. Foraging of bees on oilseed rape (Brassica napus L.) in relation to the stage of flowering of the crop and pest control. Journal of Agricultural Science, *94*, 151-154.

20. Free, J. B. and Ferguson, A. W. 1983. Foraging behaviour of honeybees on oilseed rape. Bee World, *64*, 22-24.

21. Free, J. B. and Nuttall, P. H. 1968. The pollination of oilseed rape and the behaviour of the bees on the crop. Journal of Agricultural Science, *71*, 91-94.

22. Free, J. B. and Williams, I. H. 1979. The infestation of oilseed rape by insect pests. Journal of Agricultural Science, *92*, 203-218.

23. Gardner, B. 1984. Economics of oilseed rape. Aspects of Applied Biology *6*.

24. Gontarski, H. 1957. How do bees deal with dry sugar. Deutscher Imerkalender, 149-155. (In German)

25. Hirschfelder, H. 1964. Dry sugar feeding. Deutscher Imerkalender, 75-82. (In German)

26. Kelly, F. H. C. 1954. Phase equilibrium in sugar solutions, IV. Ternary system of water-glucose-fructose. Journal of Applied Chemistry, *4*, 409-411.

27. Kramer, J. K. G. *et al* 1983. High and Low Erucic Acid Rapeseed Oils. Academic Press.

28. Langer, R. H. M. and Hill, E. D. 1982. Agricultural plants. C.U.P.

29. Lothrop, R. E. 1943. Saturation relations in aqueous solutions of some sugar mixtures with special reference to higher concentrations. Thesis, George Washington University, in Crane, E. (Ed), 1975.

30. Louveaux, J. and Vergé, J. 1952. Researches on the pollination of winter rape. Apiculteur, *96*, 15-18. (In French)

31. Mesquida, J. and Renard, M. 1982. A study of pollen dispersal by wind and the importance of anemophily in oilseed rape. Apidologie, *13*, 353-367.

32. Meyerhoff, G. 1958. Behaviour of bees foraging in rape. Bienenzeitung, *72*, 164-165. (In German)

33. Ministry of Agriculture, Fisheries and Food, 1976. Approved Products for Farmers and Growers. M.A.F.F. Reference Book 380 (76). H.M.S.O.

34. Ministry of Agriculture, Fisheries and Food, 1980. Approved Products for Farmers and Growers. M.A.F.F. Reference Book 380 (80). H.M.S.O.

35. Ministry of Agriculture, Fisheries and Food, 1982. Pesticide Science. M.A.F.F. Reference Book 252 (82). H.M.S.O.

36. Ministry of Agriculture, Fisheries and Food, 1983. Approved Products for Farmers and Growers. M.A.F.F. Reference Book 380 (83). H.M.S.O.

37. Ministry of Agriculture, Fisheries and Food, 1983. Pesticide Science. M.A.F.F. Reference Book 252 (83). H.M.S.O.

38. Ministry of Agriculture, Fisheries and Food, 1985. Approved Products for Farmers and Growers. M.A.F.F. Reference Book 380 (85). H.M.S.O.

39. Persson, B. 1956. The importance and extent of intercrossing in rape (Brassica napus L.) with especial reference to the activity of honeybees. Statens Vaxtskyddsanstalt Meddelande, *70*, 5-36. (In Swedish, English summary)

40. Ribbands, R. 1953. The behaviour and social life of honeybees. Bee Research Association Limited.

41. Röbbelen, G. and Thies, W. 1980. Biosynthesis of seed oil, in Tsunoda, S. *et al* (Eds.). Brassica crops and wild allies. Japan Scientific Societies.

42. Shell International Chemical Co., Ltd., 1985. Fastac and the environment. A publicity booklet.

43. Simpson, J. 1964. Dilution by honeybees of solid and liquid food containing sugar. Journal of Agricultural Research, *3*, 37-40.

44. Tasei, J. N. 1977. Observations in winter rape: flowering, nectar secretion, foraging by honeybees. Bulletin Technique Apicole, *4*, 9-16.

45. Timms, L. M. 1983. Forms of leg abnormality observed in male broilers fed on a diet containing 12.5 per cent rapeseed meal. Research in Veterinary Science, *35*, 182-189.

46. Wheatley, G. A. and Finch, S. 1984. Effects of oilseed rape on the status of insect pests of vegetable brassicas. Proceedings of the British Crop Protection Conference, *2*, 807-814.

47. White, J. W. 1975. Physical characteristics of honey, in Crane, E. (Ed.), 1975.

48. Williams, I. H. 1978. Pests and Pollination of Oilseed Rape Crops in England. Central Association of Beekeepers.

49. Williams, I. H. 1980. Oilseed rape and beekeeping, particularly in Britain. Bee World, *61*, 141-153.

50. Williams, I. H. 1984. Concentration of airborne rape pollen over a crop of oilseed rape. Journal of Agricultural Science, *103*, 353-357.

51. Williams, I. H. and Cook, V. A. The beekeeping potential of oilseed rape. Rothamsted Experimental Station and A.D.A.S. National Beekeeping Unit. A pamphlet.

52. Wykes, G. R. 1952. The preferences of honeybees for solutions of various sugars which occur in nectar. Journal of Experimental Biology, *29*, 511-518.

53. Zander, E. 1952. Rape and bees. Zeitschrift für Bienenforschung, *1*, 135-140. (In German)